PROCESSED REALITY

Scholarly Articles by Peter Fritz Walter

The Law of Evidence

The Restriction of National Sovereignty

Alternative Medicine and Wellness Techniques

Consciousness and Shamanism

Creative Prayer

Soul Jazz

The Ego Matter

The Star Script

The Lunar Bull

Basics of Mythology

Basics of Feng Shui

Power or Depression?

The Mythology of Narcissism

Normative Psychoanalysis

Notes on Consciousness

Patterns of Perception

Sane Child vs. Insane Society

Basics of the Science of Mind

The Secret Science

Oedipal Hero

Processed Reality

PROCESSED REALITY

Pitfalls of Perception and the Cosmic Mind

By Peter Fritz Walter

Published by Sirius-C Media Galaxy LLC

3511 Silverside Road, Suite 105, Wilmington, Delaware, USA

©2017 Peter Fritz Walter. Some rights reserved.

Creative Commons Attribution 4.0 International License

This publication may be distributed, used for an adaptation or for derivative works, also for commercial purposes, as long as the rights of the author are attributed. The attribution must be given to the best of the user's ability with the information available. Third party licenses or copyright of quoted resources are untouched by this license and remain under their own license.

The moral right of the author has been asserted

Set in Avenir Light and Trajan Pro

Designed by Peter Fritz Walter

ISBN 978-1-981933-72-3

Publishing Categories
Science / Life Sciences / Neuroscience

Publisher Contact Information
publisher@sirius-c-publishing.com
http://sirius-c-publishing.com

Author Contact Information
pfw@peterfritzwalter.com

About Dr. Peter Fritz Walter
http://peterfritzwalter.com

About the Author

Parallel to an international law career in Germany, Switzerland and the United States, Dr. Peter Fritz Walter (Pierre) focused upon fine art, cookery, astrology, musical performance, social sciences and humanities.

He started writing essays as an adolescent and received a high school award for creative writing and editorial work for the school magazine.

After finalizing his law diplomas, he graduated with an LL.M. in European Integration at Saarland University, Germany, and with a Doctor of Law title from University of Geneva, Switzerland, in 1987.

He then took courses in psychology at the University of Geneva and interviewed a number of psychotherapists in Lausanne and Geneva, Switzerland. His interest was intensified through a hypnotherapy with an Ericksonian American hypnotherapist in Lausanne. This led him to the recovery and healing of his inner child.

In 1986, he met the late French psychotherapist and child psychoanalyst Françoise Dolto (1908-1988) in Paris and interviewed her. A long correspondence followed up to their encounter which was considered by the curators of the Dolto Trust interesting enough to be published in a book alongside all of Dolto's other letter exchanges by Gallimard Publishers in Paris, in 2005.

After a second career as a corporate trainer and personal coach, Pierre retired as a full-time writer, philosopher and consultant.

His nonfiction books emphasize a systemic, holistic, cross-cultural and interdisciplinary perspective, while his fiction works and short stories focus upon education, philosophy, perennial wisdom, and the poetic formulation of an integrative worldview.

Pierre is a German-French bilingual native speaker and writes English as his 4th language after German, Latin and French. He also reads source literature for his research works in Spanish, Italian, Portuguese, and Dutch. In addition, Pierre has notions of Thai, Khmer, Chinese and Japanese.

All of Pierre's books are hand-crafted and self-published, designed by the author. Pierre publishes via his Delaware company, Sirius-C Media Galaxy LLC, and under the imprints of IPUBLICA and SCM (Sirius-C Media).

Pierre's Amazon Author Page

http://www.amazon.com/Peter-Fritz-Walter/e/B00M2QN4SU

Pierre's Blog

https://medium.com/@pierrefwalter

Contents

Introduction — 11
Can We Perceive Reality Unprocessed?

Chapter One — 21
Processing Reality

Chapter Two — 29
Pitfalls of Perception

The Memory Matrix — 30
Processed Reality — 35
- Self-Fulfilling Prophecies — 38
- Unconscious Repetition Urges — 42

Spiritual Pitfalls — 47
- Churches — 47
- Sects — 50
- Gurus — 52
- Saviors — 54

Ideological Pitfalls — 56
Emotional Pitfalls — 63

Chapter Three — 69
The Myths of Worldwide Democracy

The Myth of Child Protection — 69

THE MYTH OF CIVILIZATION	76
THE MYTH OF CONTROL	78
THE MYTH OF CULTURE	80
THE MYTH OF EDUCATION	86
THE MYTH OF MORALITY	87
THE MYTH OF NORMALCY	90
THE MYTH OF POVERTY	92
THE MYTHS OF RELIGION	95
THE MYTH OF SCIENCE	97

CHAPTER FOUR — 101
Creating Reality

BIBLIOGRAPHY — 119
Contextual Bibliography

PERSONAL NOTES — 129

All of you will need to heal, decode, rescript and reprogram, so that fear is no longer a part of your psyche.

>   —Wendy Munro, Journey into a New Millennium (1997)

Introduction

Can We Perceive Reality Unprocessed?

One of the essential questions I am asking in this article is what conditioning is, why humans have a strong tendency to be conditioned to all kinds of social rules and norms, how conditioning is effected and what purpose it serves.

As a second step, then I try to make out some hidden connections between conditioning, perception and cognition.

To begin with, I am asking upfront if it is at all possible to enjoy reality *unprocessed*, as raw information that comes in through the senses, or if reality as we perceive it is already a processed one in the sense that it is distorted by our perception interface, and therefore, conditioned? Hence, when we want to find out about the subtle techniques of cultural conditioning, we better look first into its contrary and ask: what is that state that is not conditioned. This leads us automatically to the question what reality is. Is reality a matter of perception? Or rather a matter of beliefs? How do our beliefs impact and imprint upon our perception? What makes us believe at all?

Do we need to believe anything? Or is making up beliefs perhaps an automatism of our brain?

Is reality rational, irrational or meta–rational? Do we live in different realities or is there one single monolithic reality?

Further, can we perceive reality in a pure, unprocessed state or is every reality we perceive a 'processed reality' already, namely in the sense that our perception is conditioned by the feedback–loop of our memory surface? And can we clear our perception surface to a point to perceive reality, whatever it takes to get there, in a pure and unprocessed state? Is the universe a creation of our thoughts, or is it an abstract creation? What about

CAN WE PERCEIVE REALITY UNPROCESSED?

poetic reality? How is it possible that great poets and thinkers have intuited most of the findings that quantum physics now reveals to us?

And once we have clarity about how perception works, and how easily it is manipulated by the big commercial and political players, we may inquire into that latest slogan they have come up with some years ago, and that reads worldwide democracy. And we may ask: 'What is that soup and what is the taste of it?'

What is worldwide democracy and in which ways does it infringe upon our perception of reality? And upon which myths is it built?

These are some of the questions I try to ask and answer as good as I can in this essay. By the way, those who see the solution of all problems in the propagation of what they call rational reality or logical reality are mistaken as they blind out the irrational, which is equally part of human nature. Our emotions have their own intelligence, and yet they are not rational when perceived only from the left side of the river, our deductive mind. We have a right brain hemisphere, and that functions pretty much on the irrational side most of the time, and through associative logic.

We cannot perceive the totality of reality as long as we blind out a part of our human setup. We are totally aware when we are totally human. And we are totally inhuman when we are totally rational. The intelligence of nature is not rational, but meta–rational or holistic.

Let us have an overview over 1960s and 70s neurological research and the early theory of 'preferred pathways,' proposed by neurologist Herbert James Campbell in his book The Pleasure Areas (1973), as well as the 'mechanism of mind' theory of think tank Edward de Bono. We see that this research clearly suggests that the brain 'can only see what it already knows' (Edward de Bono) and that as a result, researchers can only pick out the data from their research that confirms their basic assumptions and beliefs.

This research thus anticipated the basic tenet of quantum physics that however we set up an experiment, the observer shall always be entangled with the object he observes. In addition, this

CAN WE PERCEIVE REALITY UNPROCESSED?

mechanism of our brain, which serves survival, not the highest possible accuracy of perception, makes for fundamental novelty most of the time brought about by accidents, mistakes, lapses of attention, and the like, because as long as the mechanism of our brain is intact, we cannot experience novelty.

Does that mean that the door is closed forever in the sense that direct perception is an impossibility? Despite this setup of our perception interface, direct perception is possible, while it's possible only for a select few of shamans, yogis, zen masters and generally, people who have really worked through their emotional entanglements, and who reduced their projections and blown–up belief system to a strict minimum. As we thus can acknowledge that direct perception is possible, we can ask what the precise criteria are for it to happen? We see, based on this insight, that we are actually educated in our culture to misuse our brains and to develop a faulty perception interface.

In an attempt to define Love, Krishnamurti once said that Love cannot be defined, or it becomes a concept; thus to understand Love, we can only approach it negatively, looking at all what is not Love. This is the approach I am taking in this essay, showing with a number of present-day examples, that what is generally taken for 'reality' is a 'processed food' in the sense that it's pervaded with concepts, ideological, religious or scientific, and thus data that is not part of the original idea.

To remind one of the several examples of 'myths' I am going to present further down in this essay, child protection, we can see that originally, the idea was to protect children from harm, as simple and as effective as this sounds. But then, when we look at what 'child protection' really does, what it really leads to in our society, what it really implies, what it really results in, then we see that it creates more harm to children than the situation when children go unprotected, that it creates more confusion and more violence than any given society would experience that doesn't care, or cares to a lesser extent, at protecting their children in an organized and police-backed manner. And last not least, we see that this paradigm really interferes with children's natural sexual growth in that it demonizes and persecutes any attempt of a child to experience sexual pleasure with another child or an adult, thus pervading the early environment of children with fear, a real fear of life that can develop into hysteria and paranoia, and can in extreme cases lead to autism, learning disabilities, anorexia, depressions, epilepsy, and even schizophrenia.

Taking a kind of bird perspective, we can only wonder how a society can declare itself to being 'scientific' and 'enlightened', compared to a religiously fundamentalist regime, and nonetheless bring about a reality that is composed of exactly the same elements that fundamentalist regimes are putting in place, such as strong fear of authority, meticulous state supervision ruling into human relations, and especially, the family, large–scale prohibitions targeting primarily at human intimacy, speech taboos, large–scale censorship of 'sensitive' issues such as adult–child sexual behavior, and draconian punishments waiting for the law breaker. And we can only wonder how such a society can be so bold and daring to sell this mix of violence, state–ordained brutality, ignorance, manipulation and suppression of non–system–conform research, taboo–thinking, retrograde provincialism and fascism in a garment called 'worldwide democracy'?

PROCESSED REALITY

To remind my base intention when writing this essay, it was not meant as a pamphlet or social critique, or else a form of political activism. If this had been the cause, it would have been published in a different manner; hence my base intention was to present the data as a consciousness opener.

If such doing is possible in just any kind of society, be it based upon religious or agnostic fundamentalism, then something must be basically wrong with how we handle perception, with how we see and evaluate reality, with how we are brought up, with how we have learnt to handle reality. Then we need to actually reframe our perceptions and clear our perception interface; we might also want to take a role in the educational or political world so as to work against ignorance and for fostering real information, one namely that is based on correct perception, perception that is as direct and pure as possible, perception that is as much as possible free of contamination, of lies, of political maneuvering, and opportunism.

Hence it is of paramount importance how we handle our perception interface; from this insight, we

CAN WE PERCEIVE REALITY UNPROCESSED?

can then set about to improve self–reflective consciousness, to build awareness of our thinking habits and behavior patterns, and see how our actions impact upon society as a whole, considering karma for a whole group, or even a nation, or the entire globe. When we do this, we see that we are responsible, through the very fact of participating in creation and co–creation, or as Dr. William Tiller put it in the movie 'What the Bleep Do We Know!?, in our being 'avatars'. This, then, will shift our regard and deprive us from comforting projections such as 'our government is the culprit' or 'modern–time debauchery and lack of values is the culprit' or 'lack of religious belonging is the culprit'. We then see that the universe doesn't ask who is a culprit for anything that ever happened, but asks us to stay true to the basic principle in a responsive universe, that is, individual response–ability.

This, in turn, leads to a careful approach when meeting ourselves, in our daily self–talk, when meeting others in our daily dialogues and exchanges, and in our cooperation with others. The starting point always is how we perceive reality, and how we still better perceive reality when being meticulous, just like a warrior, in keeping our perception interface as pure and pristine as possible.

Chapter One

Processing Reality

Institutional education does not serve us. It serves the system, and not the child because it does not recognize the existence of individual, and individually gifted, children. For mass education, there is a quantity of humans to be educated, and not a variety of unique individuals.

This is the reason why mass education is destructive and leads to devolution, not to evolution. It serves not cultural, but if ever, military needs. It destroys what is human in us. It suffocates sensitivity and sensuality, wipes out individual differences by imposing standard patterns of behavior, with the ultimate purpose of molding children into a standard scheme of thinking and acting that is socially approved and politically correct. It conditions good humans to become bad citizens and dull bureaucrats. It paralyzes the inborn self-thinker and replaces individual intelligence by standard stupidity.

Even in schools for high-class children, the child is often not respected as an integrated whole but reduced into a split-self and conditioned to become a cunning career-hawk; by doing so, the unique soul qualities of the child are neglected if not shunned, and the outright focus on left-brain, separative yang values, the more integrative and socially functional yin values are wiped under the carpet. This is how, virtually in the cradle, and the crèche, hubris and violence are structurally programmed by modern society.

A further result of this educational bias is that systemic, associative, creative and ecological thinking is mutilated and so-called rational, logical, strategic thinking is hypertrophied. The outcome is the functionary, the ruthless executioner or the career woman who lets her children over to daycare and never even thought why, in the first place, she wanted to have a child. Thus, mass education is a quite effective form of child killing.

Traditional education suffocates the child's emotional life instead of understanding that child-rearing is first of all taking care of the child's healthy emotional development. Modern education is not different from traditional education in this respect; it differs only in that it castrates children emotionally, less through brutal nonsensical prohibitions, but rather through cunning intellectual dressage. The result is the same: children who prematurely excel in doing every kind of puzzle and are brilliant in small-talk while being hyperactive, bed-wetting, and insomniac. Not to mention that they lost their sexy charm and look and act like dwarf-adults.

To focus on every individual child is only possible when, from the start, we have a qualitative and not a quantitative approach to education. The quality approach does not ask for efficiency, but for integrated solutions that serve every child in the community. Life education opposes moralistic education because morality-based education serves the dominator culture, and not the child.

Dogma-based education moulds the child after the adult role model as a social ideal, and thus after the values valid for a majority. Such form of education regularly disregards the true needs of the child or sacrifices them on the altar of mighty parenthood. Moralistic education is contrary to intelligence for it creates psychological fear and guilt which in turn build up barriers to self-knowledge. Intelligence is nourished by observing our psycho-emotional reactions as well as our actions on a regular basis.

The adherence to morality or ideological positions hinders the birth of true intelligence.

Education based upon ideologies fosters negative growth, absolute rigid positions, conformity, imitation and, in last resort, violence. In order to grasp an idea of the emotional life of the child, we need to grasp what is intelligence.

Most people confuse intelligence with knowledge. They don't see that accumulation of knowledge is purely mechanical and as such no indicator of intelligence. Intelligence is something entirely different from knowledge. It is not mechanical, but a natural by-product of emotional vivacity and wholeness: to grow sanely means to be not fragmented and rather intuitive.

Children and geniuses have in common that they are emotionally intact and that they are not fragmented by moralistic trash. Our rational mind only functions at full capacity when it is connected to our irrational mind so that intellectual/analytic and intuitive/synthetic thought synergistically interact with each other. Then, regularly, the rational and the emotional part of us are well balanced and we experience a state of lasting inner peace.

Mainstream education makes a complex and difficult natural process easy by destroying it. It kills the child emotionally by invalidating the child's right-brain capacities and resulting actions and, as a matter of fact, castrating the child emotionally at an early age. The world is populated with people that are emotionally dead. Mainstream education reminds of a gardener who, because of lacking knowledge how to grow a specific plant, just roots it out. Junk child!

Life education does not have and does not need junk children as mainstream education does. Junk children are the reminder of civilization. That's exactly what they do, then: they remind us of our incapacity to educate them properly.

Drug children, violent children, hypersexual children, drop-out children—all those are variants of the junk child: those who used to stand in the corner for hours, crying, weaning, rejected.

Mainstream education shows its ignorance in how they handle junk children, in how they *produce* junk children. Instead, life education develops wholeness for all children by empathetically and holistically, and not only intellectually understanding children.

The natural counterpart of wholeness is holistic thinking. Intelligence, sensitivity and understanding for the complex functions of life can only be developed if cognition is imbedded in the emotional life of the person and thus a result of wholeness, and not of fragmentation.

Intellectual capacities and skills that have no connection with the emotional life of the person and that are disconnected from the heart are truly dangerous. They transform a human being into a functionary that is able to lead concentration camps or will click-out bombs on forest children if those actions only fit in their mental concepts. Only sensitivity acts counter to cruelty, not the cultivation of thought systems, ideals or religions.

Cognitive capacities that are imbedded in emotional healthiness can only grow on a basis of readiness, of maturity. A child will voluntarily accept instruction in reading and writing once emotionally ready for it and not under any other circumstance. And here I speak about the individual maturity of a child, and not a standard concept, since there simply are no standards. Education must logically proceed in a one-to-one relation and interaction between educator and child, for only then the uniqueness of the child can be validated. The emotional bond in this relation is of overwhelming importance. Only love can be the bridge for the transmission of values.

Group education therefore is fake-education. It seems to be more effective while truly it is much less since the child, in a group, is treated at a bottomline level and not according to their emotional and intellectual complexity. In addition, tactile stimulation is of high importance, especially for smaller children because it enhances the emotional flow and by doing so strengthens the child's immunitary system.

In a time of manifold immunity-related diseases, we should begin to appreciate the preventive benefits of tactile stimulation and consider to incorporate child massage in our educational curricula.

—See, for example, Frederick Leboyer, Loving Hands: The Traditional Art of Baby Massage (1977).

Chapter Two

Pitfalls of Perception

Edward de Bono was one of the first coaches and professional think tanks who, after thorough research into the functioning of the brain and human perception, found that *human perception is faulty in the sense that its self–organizing structure is conditioned to ensure survival*, and not to bring about the highest possible level of integrity in processing and storing the data that is coming in through the senses.

When the brain receives new information, it has only *two alternatives* for storing that information away; it builds a new pattern or adds the data on to an existing pattern. The first alternative would ensure the highest level of preserving the integrity of the incoming information. The second alternative ultimately brings about a faulty perception interface, but it's much faster to process for the brain, and it serves survival because it builds upon already established information pathways in the brain.

The Memory Matrix

When de Bono, in 1969, released his theory of passively organizing systems in one of his first books, *The Mechanism of Mind (1969)*, scientists at first disregarded his research. When later Nobel prize winners confirmed it, and amazing new discoveries in neurology corroborated it, de Bono became well-known and his advice was sought after by some of the largest corporations in the world.

The preferred–pathways theory, now presented even in popular science books, is the scientific formulation of de Bono's early theory.

Historically, de Bono certainly was the first author to write about the negative side of this system whereas many neurologists, until this day, continue to recognize but the positive effects of it. The essential negative point in self–organizing systems is that the recognition of new patterns is conditioned upon the characteristics of already existing patterns.

Bono said in his book *Serious Creativity (1996)* that when we analyze data we can only pick out the idea we already have. That de Bono's insight is more than neurology is shown by the fact that Krishnamurti, Maharshi and many yogis teach us that only *total awareness*, not thought can help us understand the world intelligently.

Thought or the *rational mind* is not able to recognize patterns; it can only process patterns that are already stored away in the memory surface. The conditioning of perception by thought and by past experience was one of the arguments Krishnamurti used to show *how to overcome the limitations inherent in the thought process*. Krishnamurti showed in his writings and talks that there is unlimited intelligence and awareness not in thought but in the realm beyond thought.

In simple language, when we perceive reality, the reality we perceive never looks fresh and new, but as something already known; this is so because the brain, as a matter of automatism, conditions the new information it receives upon what it already knows. Practically speaking, the patterns we perceive that can fit in existing patterns are automatically added–on to these existing patterns by the brain.

When it occurs that a new patterns reveals to be so different that it cannot be added on to any existing pattern, the brain, *instead of immediately building a new preferred pathway,* will try to bend the new pattern so much that it becomes similar to an existing pattern. Practical example: when you see an UFO landing and some extra–terrestrials leaving it, you will look at this in amazement, but later, in hindsight, you will tend to argue 'Oh yes, I guess it was just a normal airplane that I saw landing, and I was probably *hallucinating* altogether in that moment.' Your brain added the new information on to already existing information instead of forming a new pattern in the memory surface. This is how the brain, and the process of thought, works, and how this system impacts upon perception by actually per se distorting perception.

The British neurologist Herbert James Campbell gave comprehensive answers. Campbell argues that our brain has developed this kind of faulty memory surface because it was protecting human survival— while by doing that it has brought about billions of deficient thinkers! Now, how can we avoid this automatism? Krishnamurti taught that it was by practicing total attention. We can be so alert that we are aware of the brain's attempt to trick us out. A woman says:

> —I was once stolen money by a trickster. Today I saw a charming young trickster and stage hypnotist. We were flirting a moment. Then, when I looked in his eyes, it came to my mind that he just wanted to steal me money.

The new pattern was: A trickster—potential lover. The old pattern was: A trickster—thief. The new pattern did not fit in the memory surface. The pattern was: a trickster—thief. The new pattern a trickster—possible lover could not 'erase' the old pattern nor could it be added–on to it because of the contradiction thief–lover. What should we tell that lady regarding the old pattern? We saw that her brain could not erase the pattern automatically, but that some kind of input from herself was needed. I would have the following dialog with the client:

> —Please first question the validity of the old pattern! Was there not a logical fault in the old pattern as it was a generalization? A trickster who stole you money, yes or no?
>
> —Yes.
>
> —All tricksters steal money. Yes or no?
>
> —No.

Through raising her consciousness by these simple questions she could indeed have erased the old pattern; however, this is only valid for the brain, for her intellect, not for her heart: her emotions could still adhere to the old myth that all tricksters or stage hypnotists were thieves. So I tell her:

> —Please engage in a new love relation in order to *disprove the validity of a single experience* that was triggering a general belief. Through a new positive experience the belief you have stored away as a result of an earlier experience can be effectively erased.
>
> —Do you guarantee that? she asks.
>
> —There is no guarantee in life, regarding love, I reply. If you want a guarantee for love, you kill love.

> When you love, go for love, not for security.
> Security is the death of love.

It is obvious that the second method is better than the first, because it will impact both upon the mind and the emotions of the subject.

Processed Reality

When I have a certain opinion about books, my whole attitude toward books, my handling of books, my appreciation or depreciation of books, and my habits for purchasing books are all impregnated by my opinion about books.

Whatever made me form that opinion in the first place—most of the time I will have forgotten about it anyway—is not important. The opinions I cherish have an *immediate impact* upon my perception of reality. Instead of leaving reality as it is, the brain thus processes reality by the very mechanism of perception that it uses to perceive this reality.

When I have certain religious convictions, such as the conviction that eating pork meat is bad for my health, mind, growth and attitudes, and perhaps even for my sexual behavior, my relationship with pork meat is impregnated by this conviction: in the most common case, I will make a big circle around pork meat. When I eat pork, and as a result of my convictions, I will think that I am a swine. It works like that. And in talks with others I will stress the undesired effects of pork meat. I will certainly find a great number of scientific research that proves my point of view, and thus validates my conviction that pork meat is a harmful component in the human organism. Let's get free of swinish things and habits …, you will declare, and conclude, with your habitual enthusiasm:

> —Oh folks, begin to pray for the new religion that is free of swines!

Still stronger than opinions and convictions are *beliefs*. What are beliefs? Figure beliefs as highly condensed convictions, so condensed that they have an immediate and absolute impact on our perception, an impact that is so strong and direct that we are totally unaware of it. The danger of beliefs is in fact that in most cases they are completely unconscious.

In fact, all our life circumstances are but reflections of our inner life, projected upon the interface of *real life*. From our inner state, the screen of thought and of our conscious and unconscious beliefs, energy irradiates into the universe that brings about changes and that drives us and others to various kinds of actions. Depending upon the level of integration and harmony of our inner actors, the resulting actions are effective or ineffective, constructive or destructive, harmonious or disruptive.

That is why beliefs immediately *condition* our reality. They are very important keys to understanding our personal reality, while they are really in the way of understanding our soul reality. But beliefs lose their power, and they are no more a trap to holistic perception the moment we understand their impact, when we see how powerful they actually are, and how dangerous.

When I know my beliefs, I understand my life. When I am *unconscious* of my beliefs, I am a ball for others to play with because my life is fake, as I am lacking authenticity.

Why? Because I am lacking autonomy which is a state of consciousness that is free of limiting beliefs and idiotic convictions. Living without convictions and values is the only way to be free.

Values, the big word in American power training is in fact the greatest manipulation. Because it is *but belief*. There are no values other than what I project upon the surface of my consciousness. I may setup guiding principles in my life and call them values, but what most people call values are *beliefs*.

Self–Fulfilling Prophecies

Prophecies are not only those by famous seers such as *Nostradamus*, or those contained in the Bible's Apocalypse, but also prophecies that we receive from astrologers, fortune–tellers, or numerologists. Apart from the creative flow that marks the distance between a poetically expressed prophecy and a precise time–lined event in world history, there are no popular books that report how many of Nostradamus' predictions have *not* realized in tangible reality. Thus, the proof, if there is any, would have to be corroborated by counter–proof. This is not a matter of trust or mistrust, but a matter of statistics.

Responsible and honest astrologers, numerologists and Tarot experts do not work with prophecies, and they consciously avoid being *suggestive*. As all fortune telling is but *scanning the content of consciousness* and extrapolating it onto the future, it is a volatile thing because through changing the content of my consciousness, I change my future implicitly. That is why a responsible and conscious astrologer or diviner will tell the client only the present content of their consciousness as it appears from the planetary constellations in the birth chart and all the additional vectors such as *Moon Nodes, Transits and Progressions, Part of Fortune* as well as *Lilith* and *Progressed Lilith*.

We *cannot know or predict the future* as the future is based upon the present and the present is subject to constant change and transformation. The change of the vector *Present* triggers the vector *Future* to change accordingly. Thus every prediction of the future truly is an inquiry into the content of present consciousness. Here the paranormal element comes in, which is simply telepathy. The worst prophecies are those that we use to call *self–fulfilling prophecies*, those that we give to ourselves, as a form of voicing our beliefs. For example Mister X. loudly voices at a party that he just has started a new business, and then takes a deep look in the beer glass, after which he declares:

> —Well, well, but ... surely ... as I know myself ... and my life ... all this will end like all ended before: in a complete failure ...

And everybody laughs. While there was nothing, absolutely nothing to laugh about. Somebody killed himself in front of the whole audience. *Self–fulfilling prophecies are suicide.* And they have been shown to be involved also in the etiology of cancer. Recognized alternative cancer therapists Dr. Carl Simonton and Stephanie Simonton write in their book *Getting Well Again (1978/1992)* that a patient who avidly expects recovery has clearly more chances for complete healing than one who expects to die because he or she has given up any hope for recovery. They speak in either case of a *reinforcing cycle* that is put in motion through the *expectancy* they harbor and feed in their mind. They write:

> [A]n expectation of success will often lead to success, which in turn provides evidence / that the original expectation was correct. On the other hand, an expectation of failure will often result in an unsuccessful outcome, which in turn validates the negative expectation. In both cases, the outcome

created by the expectation supports the validity of the original expectation. The expectancy, whether positive or negative, gets stronger the more the cycle is repeated. (Id., pp. 80-81)

What is valid in medical science, the Simontons pursue, is also valid in education. They report in their book the following startling survey:

> Rosenthal and his associates produced equally startling evidence on expectancy in a study conducted with children in a California public school district. A non–verbal intelligence test was administered to eighteen classrooms of elementary students at the beginning of the school year. The teachers were told that the test would predict which children were ready to bloom intellectually. Twenty percent of the students whom Rosenthal selected at random, and not on the basis of test scores, were then identified as being 'intellectual bloomers', and their teachers were told that these students could be expected to show remarkable gains during the coming year. The only difference between these students and a control group was the expectancy created in the teachers' minds. Yet when both groups of students were retested eight / months later, the randomly chosen 'bloomers' had gained in I.Q. points over the control group.
>
> —Id., pp. 81–82. The original research can be found in R. Rosenthal, Experimenter effects in behavioral research, New York: Appleton–Century–Crofts, 1966.

Unconscious Repetition Urges

Unconscious repetition urges were first discovered by psychoanalysis. *Sigmund Freud* found a mechanism in our psyche that is quite uncanny : when we suffer a traumatic event, especially in childhood, our brain has only two possibilities:

- shut off the computer: we turn mad; or

- reprogram the software: we forget all.

The brain avoids the first, as a matter of self–protection, and practices the second. While all is stored away in our memory surface, the conflictual content becomes repressed into the unconscious. And then, it happens that we attract circumstances that lead us to again and again repeat the same scenario, really or metaphorically, and this is organized by our memory surface not for bothering us, but for giving us a chance for healing the early trauma.

Alice Miller often refers to her childhood and the traumata she has suffered from a deeply narcissistic, cold, cruel and lifeless mother. But her main problem, as probably the main problem of the patients she treats in her psychiatric practice in Switzerland, is not the hurts and traumata that she or them can remember. It is those that they *cannot* remember but that are signaling their existence through the nasty fact that over and over in their lives, they are facing the same problems with people, the same problems with partners, and the same patterns in those problems; patterns that thus are similar and repetitive, or rather one pattern that manifests with a little variety.

I have tracked this pattern and found what I call an *abuse pattern* being present in many women. For example, a mother lives within a conflictual marriage where she and her partner engage in extramarital sex and pornographic parties, something that occurs in continental Europe more often than not. She has strong fears about her children, and especially the fear that her little girl could be sexually abused by a stranger. As a result of her obsessive fear, she is suspicious about her babysitter, as he is a male, and she is kind of convinced that all males want to abuse once in a while, just to prove themselves they are males. After further talks, I invariably found with such clients that they had been abused in their younger years or during their girlhood by either their father or stepfather, their grandfather, an uncle, an older brother, a cousin, or a boyfriend when they were dating for the first time. Thus, I did not need to read Alice Miller to know about that specific pattern. I had enough crying mothers in front of me, when over a cup of coffee they told me about what they called their 'problem.'

And in a case that particularly touched me, I was dumbfounded to see how inescapable the fate seems to be that unconscious repetition urges bring about. When I was seven years old, I fell in love with a neighbor girl, Ursula.

Ursula was the daughter of a policeman who was working with the French military, the former occupation force in our town. He was thus a horse-top policeman, called in French, a *cavalier*. But he was not a gentleman toward his daughter, because he was whipping her regularly for punishment. And that girl suffered terribly from the sadism of her father and our relationship was rendered impossible by that sadist and equally by the girl's very suspicious and neurotic mother, to a point that she simply was forbidden to ride on the bike with me. She was never allowed to leave the house when I came to pick her up.

After almost twenty years I met her again, by chance, in a little bistro in our town. And she told me the following story:

> I have suffered terribly from men in my life. First my father and then my husband. You know what my father did to me, but you don't know what my husband did. He raped me constantly and was beating me, so that I asked for the divorce. We were divorced and after the divorce, he broke in my apartment and raped me again. I had to call the police but they said they could not really protect me.

Unconscious repetition urge means that destiny wanted Ursula get beyond her affliction and once for all *solve her victim condition*, a problem of co-fusion, of co-dependence puzzled up with hate-love feelings toward her father. You may think Ursula will get a gun and shoot down the next guy who attempts to take advantage of her. Yet that is not what usually happens in life. That's perhaps the movies, but not real life. And it's no solution because the abuse pattern is within Ursula, not without. And it can't be healed through violence, but only through consciousness.

It's only possible if Ursula can regress again in her childhood, during medical hypnosis or any other method that uses hypnosis implicitly, or if Ursula can play it out in a psychodrama with a male who in some way stands for all those who have abused her in the past. And through that game–like role play, Ursula could observe her reaction and her feelings and see what exactly in her attracts men that violate her integrity. When Ursula was seven years old, she could not do that work and her brain had to react with a trauma response. But with twenty–seven that archaic reaction of the brain is not any more for Ursula's best and she has to learn alternatives in behavior and regarding her expectations so as to attract men that care for her, love her and see the beauty in her.

The tragedy in Ursula's life and the lives of so many others who have had traumatic childhoods is that at least once they met their savior, but were not allowed to develop the relationship. I was for Ursula that person, that savior, the one who unconditionally loved and understood her, a boy of her age. Many dreams that I received in my childhood about her and me confirmed that.

PITFALLS OF PERCEPTION

The tragedy was that our love was impossible – rendered impossible by *exactly the violators that traumatized her entire life*. Her parents.

Spiritual Pitfalls

> The recognition of the secondary nature of the personality of whatever deity is worshipped is characteristic of most of the traditions of the world. In Christianity, Mohammedanism, and Judaism, however, the personality of the divinity is taught to be final – which makes it comparatively difficult for the members of these communions to understand how one may go beyond the limitations of their own anthropomorphic divinity. The result has been, on the one hand, a general obfuscation of the symbols, and on the other, a god–ridden bigotry such is unmatched elsewhere in the history of religion. For a discussion of the possible origin of this aberration, see Sigmund Freud, Moses and Monotheism, 1939.
>
> – Joseph Campbell, The Hero with A Thousand Faces, New York: Princeton University Press, 1949, Third Printing, 1973, pp. 258, 259, note 5.

Churches

Among spiritual sense givers churches are at the first place because they are considering themselves as *superconscious organizations* established for man's spiritual welfare – while in truth they are representations of our inner shadow, the black man that we bear inside, the part of us that is not light, not conscious, and thus in a *magic condition*, and as a result subject of, and subjected to, *myths and mythologies*.

Churches and their like institutions in other religions have the nasty habit to interfere in people's lives like your mother–in–law, telling you what you ought to do, and what you ought not to. As if you were a baby or a psychotic, or a psychotic baby. It's a fact that churches and other so–called *spiritual organizations* exert power over people that subject themselves to them as their believers. Apart from the interesting word play that this represents when we think of what beliefs are and how devastating they are for any success in life, we may understand what we can expect from churches – and what we *cannot* expect from them.

In Singapore, the Christian churches have a spicy note that distinguishes them from Christian churches in the West: when the time of the mass approaches, the yard is filled with shiny new cars, predominantly

Mercedes and BMW, Jaguar and Rolls–Royce, and inside the building the walls are covered with big and golden panels that cite in detail who has given which contribution to the church's monetary fund. The amounts range from several thousand Singapore dollars to several hundred thousand Singapore dollars each. When I asked a Christian Singapore Chinese what he thought about that, he openly laughed and declared:

> But Christian churches in Singapore have nothing to do with the Christian dogma. They are prime meeting places for business people to get to know each other and to talk business. That is why we go to Church!

And when you join a mass in Geneva's Cathedral on a Sunday morning, you will hear how spicy Swiss Calvinism can be out of the mouth of a *sadist* and world–hater that declares all pleasure as 'sin' and the children's carousel in front of the Cathedral an 'invention of the devil.'

Human theater? I agree. But what the hell has it to do with spirituality? To split life in a spiritual and a non–spiritual part per se destroys any chance to get beyond the soup of mass thinking. The way to a spiritual life simply, and importantly, begins with questioning our beliefs, and all the rest of the human soup that preaches mediocrity, fatality and misery as the human condition.

The true way to spirituality consists in learning to say *No*, and again. Until something happens that was not expected and that you might call *bliss*. The Baby Jesus was a fist in the face of all churches! And a prisoner incarcerated for loving a child is a fist in the face of all so–called human justice.

Sects

It is a tragic irony that churches are trying to distance themselves from sects. They do this after they saw that sects simply were more radical, while they are for the most part based upon exactly the same principles as churches.

When I applied for a position with the *European Parliament's Administration* in Luxembourg, I was invited to participate in a nationwide competition in which eight hundred German lawyers participated. The subject of one of the written tests, guess it, was sects. At the time, in 1983, sects had become a European problem, as not only the member states of the EU were suffering from the sect problem, but the European Union as a whole tried to get hold of the 'spiritual epidemics', so to say, by giving out guidelines to all the ministries of education of all member states about how to handle the sect problem.

What had been my recommendations? I in fact recommended to not treat the symptoms but cure the disease. What was the disease? A spiritual vacuum in young people. And I had seen how it happened with a friend of mine who was a blooming youth, artist and teacher, before he entered that sect in France, and after they had brainwashed and force–married him in that sect, he was a decrepit elder with a rigid, judgmental mindset. And that sect was named *The Church of Enlightenment.*

They obviously *and for well–founded reasons* had avoided to name the organization a sect in order to circumvent the national and supranational regulations against sects. From the basic paradigm, the life denial, the dogmatic approach, the repressive attitudes toward children, with their resulting harsh beatings, the Draconian restrictions on lifestyle, diet and sexual behavior and the arrogant preaching of so-called truth 'in the name of Jesus Christ', they were exactly behaving like most churches. And the danger for modern society comes equally from both churches and sects, because it's only the label in which they differ.

Gurus

And there are those who think they are especially modest and virtuous. They do not go to churches because they find that old-fashioned, and they do not join sects because they do not want to restrict their lifestyle. Thus, they travel to India to see Sai Baba or any other guru with a certain reputation. And then they come back with longer hair, dirty clothes and an enlightened mind.

The guru and the disciple engage in a *shared form of corruption* and they are both responsible for it, not only one of them. They both play theatre, not only the guru. They both share in the same comedy, or tragedy, as you wish to see it. They are actors on the same stage. And their credo is: you have to follow a better one so that you will become better. You have to follow a higher one so that you become higher. You have to follow a more intelligent one so that you become more intelligent. You have to follow an humbler one so that you become more humble. You have to follow a famous one so that one day you will be famous. And so on.

Guruism is one of many outflows of the *hero paradigm* that is in turn a direct result of the patriarchal rut with its rigid senior–junior hierarchy among males, and even among females, but stronger rooted among males. Gurus, not those who are *meta–teachers* by living their truth, such as Krishnamurti, but *self–declared gurus* and especially those who call themselves 'spiritual teachers' are professing to know better than their disciples how the latter should live their lives.

Where they take this superiority from remains occult, and is not explainable simply because in matters of our individual quest for truth, there is no hierarchy, absolutely no hierarchy among humans.

This is simply so. We are called upon as individuals to find our personal reality, as an intimate quest, a search for our own *Holy Grail* that can be found only in our heart, and not in the bosom of any guru.

SAVIORS

Every religion cherishes a savior, some kind of super–human that is put as the living ideal, the first–class vintage, so to say, the best of the best in terms of human perfection – or whatever that might be. And that is exactly where the problem is resulting from. Saviors are mythical figures, full of mystery and disdain, it seems, for so–called 'ordinary reality.' They are nice as actors on the stage of fairy tales. But they are not nice when taken as ideal humans, as targets for projecting our wishes to be better than we are. Because in this quality, they are truly destructive for our growth.

Saviors are supposed to live special lives, lives of magic and wonder, lives that we all would like to live. But when we look and compare, for example historical facts and feats from the real lives of Jesus of Nazareth or Gautama Buddha, we see that there is really a large gap, to say the least, between what we can find was their real life, and what later was made out of that life and out of that person.

Jesus and Buddha were later declared to be cosmic masterminds. Gee. Oh. Dee. What is Gee Oh Dee?

A new song? An old soup? The name of a famous actress? No. The name of that super–savior mastermind, the big brother robot that is supposed to be the lonely baboon, the biggest boss or the most original hacker in our universe. A story for youngsters —at best.

All savior stories are hero stories. Food not for thought, but for your inner child. When you are conscious of the damage these kindergarten stories do and have done in the course of humanity, you can prevent that the damage is going to affect you, and you can avoid to fall in the savior–trap, as a result of a *narcissistic fixation*, in becoming the savior for your mother, your father, your brother, your sister, your spouse or your child. Not only mental hospitals are full of saviors and Jesus Christs, but also most families.

Françoise Dolto, in a workshop for child therapy back in 1988 reports the case of a young boy suffering from a *priapus*, a painful long-term erection of his penis that lasted for months. The father, a bank director, was in high sorrow about what he called the 'indecent condition' of his child and took the boy to Françoise Dolto's famous psychoanalytic practice in Paris. Dolto was able to cure the boy from the long-term erection of his penis, but two weeks after the successful termination of the cure, the mother of the boy called Dolto and said her husband had died from a heart attack.

The boy had been the savior of his father and as long as he was the suffering agent, the father could keep the balance in an otherwise unbalanced life. When the boy was cured, the father's crutches were taken away and he fell struck dead.

Ideological Pitfalls

Today, the propaganda machine of *postmodern international consumerism* is the result of more than a millennium of knowledge prohibition, spiritual manipulation and systematically bred life-hate. It is the newest and the most fashionable of our present-day sociopolitical ideologies.

Consumerism blindfolds the masses surely in a sweeter way than the Church's brutal knowledge prohibition that was enforced by the Inquisition and endless witch hunts. The sweetness and big promise of freedom as an integral part of consumer culture guarantees much higher effectiveness than coercion or brute force in repressing the original love wishes of the populace.

This is so because in fact subtle media–based manipulation is used to innocuously replacing emotional and sexual longings by material longings for acquiring and possessing consumer goods. And when looking at it in a superficial way, I gain freedom when I exchange a part of my money in the bank with a nicely designed and smoothly running sports car, plasma television or computer notebook. I gain creative freedom, can move around in a splendid way, comfortably, and can get a self–esteem boost through the admiration I receive from others for being an XY–Limousine owner or a VZ–Notebook owner. Besides that, I can write, compose, draw and publish my media productions in fully using the creative possibilities of my notebook.

This superficial observation however veils the fact that the overwhelming majority of limousine and notebook owners are *not more substantially creative* after buying these goods than they were before. Or, put differently, the truth is that my real freedom is not that I have money in the bank that allows me to buy or not buy consumer goods. My real freedom is to be able to live without working for anybody, without being a slave for eight hours a day, forty hours a week. By the same token, my real freedom is my urge to be creative, constantly creative, and it makes no substantial difference if I use a notebook or write by hand on a simple sheet of paper, else paint on a canvas, or improvise on the piano.

My real freedom thus is my spiritual drive or soul's desire to surpass my mere physical human condition and express my ideas, or create art.

When you take a deeper look, you see that the masses are cheated twice through the promises of consumerism. They for the most part do not have enough money in the bank, or property, or company shares allowing them to live a meaningful existence without working for somebody. Thus, what indeed most people need and desire is freedom!

Consumer culture promises this freedom, but let us see what it boils down to. For acquiring the goods that will grant me some form of creative freedom, I need money. For acquiring the money that I need to buy these goods, I need to work, and *work more than usual*, because I need a surplus of money, a lot more than the amount I anyway need for food, clothes and shelter, or the education of my children. How to get this amount of surplus money?

Through effort, increased effort and still more increased effort. Thus, in order to *expand myself more* within consumer culture, I in fact need to *curb myself more*. For every dollar of surplus money, I need to work relatively harder than for the same dollar of money covering my basic human needs for food and shelter.

Of course, this reality is blinded out by publicity, and one essential task of publicity is precisely to blindfold the consumer, suggesting that the acquisition of goods is smooth and easy, using consumer credit as the ultimate backdoor for climbing on the bandwagon. However, it has been shown by leading financial consultants that the consumer credit is the single most destructive form of credit there is in the economy. This is so because the *consumer credit*, contrary to the business credit, is not backed up by an increase in productivity, but in the contrary needs a constant almost super-human effort to produce the surplus of financial resources needed to pay back the credit. It is a time bomb!

This is so much the more embarrassing when the duration of the credit surpasses the average life-span of the consumer good. For example, when my computer notebook is trash after two years because of the swindling progress in IT technology, and I need four years of credit duration to finance it, I will use two years of my life to pay for something that I do not use any more and that does not give me any more freedom or creativity value. Of course, you may still use your notebook, but you are highly incited to sell it for a very low price and buy a new model using another consumer credit.

This is how in very credit–intensive consumer cultures such as the United States, people who seem well–to–do on first sight, actually are often caught in a net of credits that, if the slightest event happens that disturbs the credit payback–cycle, the card house crashes and people who were enjoying life yesterday, today plan to suicide themselves.

It is not a Marxist idea, then, to say that this system incarcerates the ignorant masses in pretty much the same false beliefs than previously the religious caste did with their endless taboos. Marx and Engels have brilliantly analyzed the destructive effects of capitalism, but acting on these insights brought about governmental regulation, prohibition and coercion, unfreedom, persecution and even torture and murder. And all this justified by the initial intent to prevent the masses from indebting themselves destructively.

That's how it goes. These policies were wrong and ineffective because they disregarded the basic human need for *distinction*, and the readiness of human beings to give sustained effort, and even sacrifice themselves for a higher quality of life, a higher social status, academic or scientific distinction, opportunity for travel, more popularity or a greater circle of friends and acquaintances. Clearly, these values are not commercial and what the communist ideologies have overlooked is that the human being does not per se strive for money or for accumulating money or other riches.

What people strive for is the creative freedom that wealth brings, and they do so with very good reasons, with the main reason namely that this freedom is real freedom and not the fake freedom of consumerism. Modern culture has no idea of how to live a happy life. It replaces true happiness with fake–values that suggest consumer satiation being the ultimate enjoyment in life.

> 'I have a new car, refrigerator and air–condition, a home theater and the walls in my garage covered with books. I am a well–red and cultured person. I work hard and go to the gym in my free time. I am married with a wife and three children. My wife works. My children go to school. I am a happy citizen. I have a family. I have values. I have convictions. I have possessions. I have.'

Possessions–based lifestyle kills every form of culture. And when you look at modern life you realize that it's not a culture, but a *fake–culture*, that it's not based upon values, but upon *fake–values*, and that the people on the stage are not humans, but marionettes.

Postmodern international consumer culture is based upon the genocide of countless tribal populations who *really* knew what happiness is about and who lived happy lives – until they were massacred by value–based and principle–ridden modern citizens.

Emotional Pitfalls

> 'A random telephone survey of 800 American adults in September 1996 found that 74 percent — virtually three out of four citizens—believe that the U.S. government regularly engages in conspirational and clandestine operations.'
>
> Robert A. Wilson, Everything Is Under Control: Conspiracies, Cults and Cover–Ups, New York: Harper & Row (Harper–Resource), 1998.

What did Gandhi mean when he said that spiritual training should be *educating the heart?* Gandhi was talking about raising *emotional intelligence* and holistic thinking. *Worldwide Democracy*, as we all know, moves rather in the opposite direction.

PROCESSED REALITY

We will always have secret services, conspiracies and persecutions as long as we uphold the paranoid view that there is a *greater force* that shapes our individual destinies instead of realizing that we are *the only makers of our personal universe.*

Countless individuals throughout human history have demonstrated that they were able to shape their destinies according to their own intrinsic life paradigm and soul values, refusing to subordinate their vision under shallow mainstream convictions.

Worldwide Democracy is based upon the exact opposite vision of the human being, namely on the myth that *total consumption* is what basically makes happy living. This goal, evidently serving economic interests, is largely veiled by a moralistic and fundamentalist cover paradigm and by a bulk of *myths*. The cover paradigm, which is one of projection and persecution, is generally hidden to mainstream consumers.

The veil however can be lifted quite easily by those with an inquisitive and critical mind. The information, while it is filtered by mainstream media, is easily available through alternative presses and the

Internet. But as it does not fit in the rosy foam of glorious consumerism, thus after filtering, what remains in the mainstream media are the myths, and the projections.

The main targets of the strategy of repression–and–projection are the *Muslims* and the *Pedophiles*; both groups, strategically demonized in public with a growing emphasis on their *difference*, are supposed to embody the evil that is believed to erode the happy consumer world with its asexual and emotionally castrated children that, consequently, need to be strongly protected and meticulously supervised. This is why *child protection* is actually an over–arching myth that serves as an ideological goal and polemics maker just in the same way as the racial purification theories of an Adolf Hitler or the intellectual purification theories of a Pol Pot.

What stands behind this new form of fascism, that for this time in history truly is worldwide, as it is sponsored by the publishing multimedia giants, is a basic *denial of complexity*, which has been proven by historians, and among them especially Jacob Burckhardt, as one of the roots of fascism.

This can easily be made out by an intelligent observer, when following up the daily news on the subject of Islam and what it pretendedly is up to, the wars in Iraq and Afghanistan, and how they are justified, the new impending war with Iran, and what is advanced in public by the United States government to justify it, and the blunt ignorance and one-sided polemics that stares out of any media coverage on the subject of adult-child erotic relations.

It is not difficult to see the obvious parallels between the present scapegoat groups and the historic ones. Even today, the old persecutions have not ceased either. Anti-Semitism is on the rise again, and that despite the protection that the state of Israel enjoys by the United States of America and other Western powers. The stress when analyzing fascism should perhaps not be put so much on the *objects* of the current or past persecutions, but on the *general climate* that leads to intolerance and persecution. As to this general climate, I am certainly not alone in saying that we are again in an era of political and social intolerance, of irrationality, of blunt media manipulation and of persecution. And what I am saying here is essentially that the new salvational construct that is put in the formula *worldwide democracy* is actually at the basis of the new danger to true humanity, as it serves as an easy eye–catcher and an attractive packaging device. It's truly a Pandora box. And contained in it, there is a time–bomb that ticks.

The myths I am talking about here are observable phenomena, not only theorems, and have a rather widespread if not devastating influence upon the non–reflective and consumption–prone mass mind.

And when this is true for adults, how much more children are affected by this ultimate postmodern form of mass hypnosis!

Out of a large number of myths, I will select a few and shortly present them in this last part of the theoretical first part of the study.

Chapter Three

The Myths of Worldwide Democracy

The Myth of Child Protection

Many parents who think they are modern and generous to their children are in reality consume-training their children and molding them into co-dependent partners. Unconsciously such parents act as the long arm of political systems and ideologies subtly hypnotizing their children with the concepts they have themselves been fed with. It is for this reason revolutionary, if not considered subversive, to rear children in truth and autonomy. For such kind of education is not compatible with the Oedipal-paranoid worldview that mainstream education is based upon. To raise children responsibly does surely not mean to charge them with a burden of responsibility that they cannot meet. However, the contrary is perhaps worse.

To infantilize children, discard them out of real life, institutionalize them into plastic worlds of crap and lies, and degrade them into obligatory play is part of the tactics of Oedipal confusion that justifies its utterly manipulative attitude with the claim it was serving to protect the child. Oedipal confusion brings about highly adapted standard citizens that are deeply disloyal! In fact, the reigning Oedipal mainstream culture is a community of secret anarchists that obediently say their credo, but silently sabotage the very content of it. By contrast, education toward autonomy is based upon recognizing the existence of soul values and the unique truth of every single child, also and especially if this individual truth is contrary to the reigning sociopolitical ideologies.

It is disturbing today's global consumer culture that the child be a complete human and thus a sexual being from birth, and that, as a result, children own a birthright to have their emotions and sexual feelings respected.

Françoise Dolto, the late French child therapist, wrote in her book La Cause des Enfants (1985) that it scandalizes most parents that a child be their equal and that, therefore, most parents raise their children as formerly princes ruled their kingdom.

Why this is so is obvious: a body–conscious child is not an easy consumer of compensatory toys and a thousand devices artificially created by international consumer reality and that are after all but cultural trash.

For those who contradict this view, I recall that the repression of the child's sexuality has exactly started with the onset of the Western industrial bourgeoisies, toward the end of the 17th century, and not, as many researchers wrongly believe, with the beginning of patriarchy. See, for example Françoise Dolto, La Cause des Enfants (1985) who reports, citing Ariès, about the childhood of French King Louis XIII:

> Until he was six years old, adults behaved with the prince in a perverse way: they played with his penis, allowed him to play with their genitals and to sleep with them and play 'the little devil' with them. All this was allowed. But suddenly when he was six years old, they dressed him like an adult and he

> had to follow the royal etiquette (citing Ariès, L'Enfant et la vie familiale sous l'Ancien Régime, p. 145). Despite the trauma that could follow, he had nonetheless kept something essential since, during the first years of his life, he could live his sexuality with other adults than his parents. He had here more chance in spite of the precocious adult clothing they put him in. His example is only valid for the rich classes. However, in other levels of society, how could a child of that time repress his incestuous desire and sublimate it? (Translation Mine)

Historical studies about child rearing practices in Europe stress the fact that still in the Renaissance, the sexuality of the child was not interfered with. Back in the Middle-Ages, apart from orthodox Christian circles, it was completely free.

I would like to introduce here a useful dichotomy coined by the psychoanalyst Erich Fromm. He created a simple metaphor calling current consumer society a state of to have, and original unspoiled being, a state of to be. Our original body pleasure and natural psychosomatic connection between body and mind is the state Fromm called To Be. Consumerist industrialization replaced that condition by To Have, a state of affairs where body and mind are split and the mind acting, most of the time, against the body. This state of To Have is the primary condition for consumer society to function because if people firmly decided to stay with To Be, our marketing system would not work.

It's working because people, already in early childhood, are conditioned away from the body and into the mind, something our schools know to do brilliantly and which is emosexual castration combined with intellectual hypertrophy. In good English, these kids talk like university professors when invited at a birthday party, but their emotions and their genitals are dried out. They have turned sadistic in their general mindset, and their high–pitched voices express their mix of subordination combined with inner revolt and growing violence.

Have you ever met an emotionally and sexually sane child, a child who is sexually active? Such a child has a dark–pitched and slightly smoky voice!

Consumer culture is founded not on pleasure, but on ersatz pleasure. Ersatz pleasure is the pleasure that replaces original body pleasure; thus first of all the industrial toy. While the self–made toy still has some

PROCESSED REALITY

connection with the body, the industrially produced toy is completely alien to the child's body.

Typically this toy – which in the meantime is produced by a gigantic global industry – consists of materials not akin to the human body, such as plastic or metal. Both plastic and metal are cold and rigid while the body is warm and pliable. Unconsciously children are conditioned upon the characteristics of the toys they is playing with.

> —Be plastic! translates into 'Be without true feelings and artificial!'
>
> —Be metal! translates into 'Be hard and mechanical!'

This is how the child is molded upon the values of the culture he is born into. In addition, consumer education uses techniques of confusion as educational methods to gradually alienate the child from their own truth—which is their body continuum. The child namely thinks from the body toward the mind, and thus inductively, while the conditioned adult thinks from the mind toward the body, and thus deductively.

This means that the child's truth is defined and experienced as the truth of their body. Every truth that disregards this body or tries to set it aside will not be regarded by the child as truth. It is for this reason that children cannot comprehend moralistic educational concepts since those concepts starve the body and hypertrophy the intellect. The consequence are lifelong giant water–headed babies in the form of adults who have never made the cut with their childhood, remaining emotionally and sexually immature. True virgins. While life has not made us to remain virgins, but to leave virginity and grow into loving copulation.

The fundamental conditioning of man is accomplished at the age of six, which is since Freud an established assumption in psychoanalysis. What comes later is only polish. The Oedipal confusion cheats about this truth. It creates a confused mind within an immature and rigid nonsexual body that has lost its natural intelligence. This is neurosis programmed into culture! Oedipal confusion plays the game of eternal maternity until the baby is far older than thirty!

It loves naive mother dependence and shuns and mistreats children who are precociously mature. It blows the child care industry up to a gigantic worldwide business with children as their products!

Children who resist the cultural castration and maintain their natural capacity for sexual love and sensual pleasure are put in the corner and labeled as sexualized and delinquent. If they still dare to play their own game, the child psychiatrist is ready to interfere and to issue a certificate which will mark the social death blow: schizophrenic or epileptic.

The Myth of Civilization

Many religions have tried to force peace upon man by dogma, prohibitions and punishment. Clerical and worldly forces have imprisoned humanity in a set of tight rules, laws and prescriptions that have had only one result: to render man a violent beast full of contempt, rebellion, strife, falseness and turmoil. To get out of this net of obligations and the feeling of oppression that goes along with it, man is caught in an endless pursuit of pleasure.

To make it worse, through the split in man's mental and emotional setup as a result of the schizoid dualism that judging our emotions in good and bad ones brings about, our psyche is divided in a conscious or official part and an unconscious or unofficial one. Through the process of so-called civilization and primarily the school system with its mass indoctrination and the disregard of the individual as a unique soul-being, humanity has in fact devoluted since the great Minoan and other pre-patriarchal cultures of Antiquity.

In fact, evolution has made its way only in the tiny range of technological advancement while in all other areas of life, we are today more barbarian than five thousand years ago.

Therefore, the solution for world peace is entirely different from what clerical and worldly powers have ever taught us. Truly, only those who were considered as heretics, saints or prophets have told the truth. Buddha, when he was alive, found truth by human struggle and suffering, while after his death his teaching was perverted into its exact contrary.

Through levitating the man Buddha into a god-like tower of virtue, the applicability of his teachings for us was impeded, if not rendered totally vain.

The Myth of Control

In killing the pleasure function and submitting it to group supervision and control, the human race has signed a contract with the devil, after having created this devil as a split self that controls the controller. Most of us are dulled and pacified into aloofness, and our once critical sense has been castrated and sanitized, first through the emotional death we suffered in childhood, second through the intellectual and neurotic overdrive we are in, as adults, as a compensation for the connectedness we lost, and third through the constant mass media manipulation we are too weak or too lazy to resist.

It is a deep fatal error to think that this was a modern phenomenon.

The analysis of human evolution shows that the present worldwide murder culture has its roots in our five thousand years of patriarchal history. In fact, it is ignorant to believe anything in nature or in the human-created world could come about in a vacuum. The present multi-faceted and institutionalized murders are based upon a murder tradition that goes back to the Code of Hammurabi, or what Wilhelm Reich called The First Human No to the endless flow of life, represented by the natural ebb-and-flow of pleasure. It is equally naïve and ignorant to believe that the present catastrophic state of affairs could have come about through any specific religion or ideology.

While still decades ago, the human masses were so ignorant to believe that most ideologies were necessary or even god-given, many today start to question this. But instead of seeing the failure of all ideologies for the advancement of humanity, they blame particular ideologies, such as Islam, while upholding others, more akin to their race or mindset.

The Bible is very explicit in favoring the in-group and excluding the out-group, and murder was not only permitted but even ordained by Yahweh, the cultural divinity, when there was a need to advance in one's business through domination and the ruthless holocaust of those that were naturally opposing such dominion.

Today's short-sighted Western credo is that by and large Islamic culture was to blame for the terrible disorder in the world. This assumption overlooks that Islam shares with Christianity and Judaism the same base paradigm and murderous ideology that promises human advancement based upon the rape and repression of nature, oligarchic control and the hypertrophy of yang to the detriment of yin. This terrible lie, that is by itself the root of murder, has been paid with innumerable victims, in the past, and today, and it has caused humanity to retrograde in its evolution.

THE MYTH OF CULTURE

Freud reasoned that culture was based upon the sublimation of our instincts. While Freud clearly said that sublimation is not to be equated with repression, at the end of the day the difference between sublimation and repression seems minimal.

Sigmund Freud meant that while he was in favor of children living freely their sexuality, he thought that for cultural reasons, medical and psychiatric experts had to accept the cultural choice of the repression of the child's early sexuality, limiting their professional role to healing the neuroses and psychoses that result from this choice. A deeper look reveals that it is nothing less than an act of castration to forbid the child his or her own sexuality. This is a form of parental or educational child mutilation, a form of societal child rape, if our hypocrite culture considers it as such or not.

Since Havelock Ellis it is clear that all sexual deviations and neuroses are born out of the repression of children's and adolescents' healthy sexuality. If parents, as a consequence of their own sex–denying upbringing, are inhibited from talking openly with their children about sex, they should at least be able to psychologically support their children by adopting a permissive and non–obtrusive attitude regarding their children's intimate lives.

It is a relict of patriarchal tradition that boys and girls are treated as two different species with regard to their sexual urges and behavior. Gender impregnation that is too early and too strong will suppress certain valuable characteristics in children that are attributed to the opposite sex, as for example tenderness, caring behavior, altruism, patience with men and activity, healthy egoism and competitive thinking or impatience with women.

Havelock Ellis found that the early repression of sexuality in girls was a major factor in female frigidity. The feminist view underlines that the problem is not the repression of basic drives but the general oppression of women and children. Who can be interested in perpetuating sexual dysfunctions in both sexes as a result of sexual repression during childhood and adolescence?

Who can be interested in rising sexual deviance, suicide rates and rape? Who wants to favor the escalation of violence, racism, war and destruction? For all this is the price we have to pay for maintaining the clean, pure and asexual façade behind which we hide all the negative secondary impulses which come up as a result of repressing the natural sexual function.

It is not surprising that after centuries of sexual repression and the distortion of the knowledge about natural life functions now society needs scientific corroboration of the most banal realities of life. But after all, it's absurd. Experience shows that people with natural attitudes toward sexuality will also be inclined to accept new permissive forms of education whereas people with sexual inhibitions and moralistic concepts tend to be suspicious of new and non–authoritarian concepts of education.

So–called child sexual abuse is a fake cause, a propagandistic cover–up of the real abuse in our culture, which is rampant parent–child codependence, which constitutes emotional abuse, and besides, physical abuse. Our mass media debates regarding children's alleged sexual innocence are but fights about words; all concern about trauma through early sexual experiences are but projections of adults' own orgasmic fears and armoring against natural biological functions.

Anna Freud's research on war children in London during World War II provided striking counter–evidence to the fake trauma theories of today's mainstream abuse culture; children are not traumatized by sex, and not even by bombs falling all around of them in wars and civil wars. They are only when their parents are bundles of walking paranoia and when they thus have been conditioned to be apprehensive to all and everything in life. Natural children are fearless.

The most devastating effects in adult–child sexual encounters do not result from sex but from fear and panic associated with engaging in a tabooed and not coded form of conduct. An analogy to this situation can be seen in the psychological reasons for drowning accidents. While infants can swim without having learnt swimming, older children and adults lost this ability because of the fear associated with drowning. Research has shown that many people who died through drowning could have been saved if they had received proper psychological training to cope with panic.

Adults' fear of water is irrational in much the same way as is orgasmic fear. Water, in the subconscious, is being associated with emotions and sexuality.

Society is for a large part responsible for the killing of children in chaotic sexual encounters because of its refusal to code certain forms of behavior. This is a collective form of irresponsibility that hits society at large in much the same way, or even more, than the individual perpetrator. Moral wars and hysteria cannot replace responsible laws and rules of conduct; in the contrary, they tend to prevent or disable such rules.

This is one of the most obvious reasons why pedoemotions and the whole spectrum of sexual behavior between adults and minors needs to be coded socially.

A social code—which is much more than a legal statute in that it encompasses certain forms of conduct that are socially acceptable—is the only way to maintain culture while the present irresponsible attitude produces chaos, confusion, insecurity and, at worst, a new form of civil war.

It has been shown by a great amount of research that there is a functional link between the repression of the human sexual drive and the upsurge of violence. It is first of all the repression of the child's natural sexual function and the social disapproval of tactile pleasure for certain age groups that prepares the ground for societal violence.

Thus the cultural choice that Freud wanted to respect and preserve was a culture not founded upon nature, but upon perverted nature. And that is why all the problems that culturally today we are dealing with, first of all the problem of rampant domestic, social and structural violence, are our own creation, and not at all an outcome of any fault in the original human setup. It's our collective denial to assume responsibility for our emotions, and our sexual function, that among other factors are the most obvious pitfalls in our constant drawbacks and individual and social catastrophes.

The Myth of Education

Striving for autonomy is an inherent component in every young life. Therefore hyper–protective education clearly is a form of child abuse because it smashes children's natural autonomy through emosexual deprivation. Mainstream education sacrifices the child–as–a–person for bringing about the child–as–a–consumer. I call this education death education since it is based upon the emosexual murder of children in order to subvert them into consumer robots. To achieve this, mainstream education uses what I call braincut to castrate children emotionally and sexually.

Modern society acts here much after the example of the Church. The Church, while paying lip service to love, destroyed love by instilling in their believers an horrible fear of pleasure and by punishing, through the Inquisition, free unregulated love in every possible way, including putting people to death by persecution, torture and planned murder and genocide.

Modern society castrates the child emotionally through a hypertrophy of the left brain hemisphere and logical thinking, while downplaying, circumventing and outright neglecting the development of the qualities of the right brain hemisphere, such as creative intuitions, associative thinking, systemic thinking, fantasy and creativity.

The Myth of Morality

There is no god, no savior and no punishment. There are no wrong acts, nor right acts. There is karma only, feedback given by the universe. By observing that feedback and recognizing its nature, positive or

PROCESSED REALITY

negative, I can evaluate the nature of my actions. There is no other way. You can't do that by thinking about your behavior. Thought is circular and inbound within my own continuum. I cannot abstract from my thought and become an observer–thinker, despite the fact that great sages such as Krishnamurti told us we could develop this ability. Let's assume I have not reached that stage of development and thus am still caught in the ego–based structure. Then I have the option to observe the nature of my actions by evaluating the feedback they create in the universe.

In being careful and observing what happens around you before you take any major action, you can avoid fatal mistakes and setbacks and act in accordance with the steering power of the universe. This power is of a higher intelligence, and considers not only your actions but the actions of all other humans, of all other beings, and even the actions of natural forces. How does a particular action you are going to take fit in the universe? What kind of waves will it create? What kind of responses will it trigger? All this can be evaluated before the action is taken. And the I Ching has been created exactly to assist us in that quest.

Once you understand this, you will agree that to take blind actions is a foolish thing. And yet, most people, especially in the modern world, take blind actions all the time, and even think that it was normal human behavior. It is ignorant human behavior.

Educating children to take blind actions is irresponsible education, or no education at all. Most Western people will reply that it was through a set of firm behavior rules, so-called morally correct behavior, that positive karma could be created. However, moral correctness is on the same line as political correctness. It is a total fake as moral rules change from country to country and in some countries even from village to village, and they change over time as well, and they change fundamentally when economic conditions change.

Moralism is a fiction and one of the most sordid ways to blindfold the masses of ignorant citizens and keep them from educating themselves about the universal laws and rules that really regulate action and reaction. How much morally correct behavior has triggered wars and genocide! How many massacres have been committed in the name of well-sounding moralistic slogans, how many millions of people killed for politically and morally correct principles! That the Western world, today, is caught in a cancerous and destructive death cycle is primarily the result of several millennia of moralistic tradition and education!

The Myth of Normalcy

Normalcy does not exist in nature; it is a left–brain concept, a pure intellectual construct. The assumption that normalcy equals heterosexuality follows thus the same bias; heterosexuality as such is a concept, but nothing real. Sexuality is not a fixated condition; humans are not animals that unconsciously follow instinct and conditioning; human sexuality is not a drive – and here I am consciously contradicting Freud and sexology.

Human sexuality is not distinct, not abstract from human emotions; both emotions and sexuality are linked in that sex attraction follows emotional attraction, not vice versa. In addition human sexual attraction is not something that can be split off from the individual person; it is part of the soul level of the person and as such in itself invested with soul, with life, with vital energy. Not all in sexuality is teleological with the ultimate goal of (total) penetration; all is here subject to dialogue, to mutual bargain and discussion, to trial–and–error over time, to play–like fun, and so on. And the ultimate satisfaction is by the same token not always and I would even say typically not the sexual satisfaction as such but the emotional satisfaction or the congruence between emotional and sexual fulfillment with one and the same partner.

This can well happen between an adult and a child, despite the obvious divergence in body size and genitals. I would even say that the very fact of the existence of childlove shows that nature has not programmed us like machines that do sex in a wheel–like manner or in a robotic obsession so that all fits in each other. I once said it in a poetic manner:

—We are not machines, we are human beings. We are more like cigars, hand–made and individually differently shaded, some coming with broken leaves, some having a different tint. We are not like cigarettes, machine–produced, every single cigarette like any other one, exactly the same. There is no straight line in nature, only in human intellectualism.

Today's idea of sex is after all a sort of compulsion to do sex, which is a left–brain concept, a pure intellectual compulsion and it came up probably as an anti–reaction against patriarchal sex repression. It also is part of the competition culture where sex is most of the time a matter of performance which makes that many men are driven, time and again, into temporary, sporadic or even long–term impotence.

The Myth of Poverty

One does not need to know about differential calculus to see that we can never fight poverty as long as our national budgets allocate for military and defense purposes more than ten times the amount spent for education.

Instead of admitting the simple socioeconomic and political facts of life, most politicians try to veil this truth and instead come up with ideologies.

We had communism around long enough to see what it does with the human potential. And we have capitalism still around—for how long?

Neither the capitalist nor the communist model have found the way to sustain human potential and creative living in a non–destructive way. While the capitalist model blindfolds about the strangling effects of consumer credits, the communist model seems to totally disregard the human nature. This is why, I think, the capitalist model is still around, with all its inherent pitfalls of course, whereas the communist model has almost totally disappeared from the globe.

This being said, and positively put, it is well possible to end poverty by a deliberate global collaboration of responsible governments who, aided by private foundations and businesses, set out to really build a global community, a global market and a decent living for all peoples on the globe. It is well possible to reduce military expenses and it is well possible to dedicate a substantial share of our national budgets to the fight against poverty. But to get there, a will is needed—and this will is lacking. This is the reality and not what the media tell us as the mouth pieces of the status quo. The problem is that this will is lacking, both on the individual and the collective level. Those who are rich think that by sharing a part of their riches with those who have nothing, they will have less, they will be deprived, they will starve. This is nonsense. The contrary will be true.

If we can realize a truly global market, not a fake global market as we have it today, because in our global economy today less than twenty percent of the world population can participate, we will have much more, and in addition we'll have this much more with a much better conscience, with a much better feeling. The benefits cannot be seen in advance because growth is non–linear.

It cannot be grasped by linear, mathematical thought. In reality, once poverty is eliminated and all peoples on the globe able to participate in a truly global market, the synergy that is created through such a change would be immense, it would be so immense that no economics guru can ever adequately predict it.

And not only would we be so much richer, but our general life quality would be much higher because of the positive vibrations that this would create.

This is so because when human suffering is effectively reduced, positive and growth–fostering vibrations would virtually transform our lives and our environment. And war would be eliminated as well because war is well also a function of poverty, material poverty and poverty of comprehension typically going hand in hand.

Thus, whatever you think about poverty, you should at least be honest and when others come up with the same lies that are spread out in the media every day more, you should stand up and say :

—No. I do not want to share. That's all. Period.

Then, at least you would be consistent.

The Myths of Religion

We are today facing a terrible violent mess called spirituality, in the West and in the East. What is the most obvious in this chaotic situation is the dishonesty that sells as spiritual or religious what is but a matter of confusion and false beliefs, or else a cunning poker of socioeconomic interests played out in the name of fundamentalism. We have to face it, without moving, and then only, if we can stand it without falling in the next horror trip, into the next depression or the next alcohol or mescaline trip—we can make an evolution. That is how it is.

We are progressing despite all. You get the picture intellectually, but that is not it. The only solution is the heart, not the mind. Your knowledge from books does generally not help you much. What helps you is to gain awareness about yourself. But this quest to be yourself is impossible if you bury your feelings under a masquerade of—

—How should I be?

—How should I behave?

—How am I being adequate?

—How should I plan my future?

—How am I to grow?

—How am I to develop my full potential?

—How is life and what is it for me?

—What is the sense of all living?

—What is the sense of love?

—Why do we have sexual urges?

It's only after stopping to ask and accept life as it is and yourself as you are, and your being imperfect as a good and normal condition, that you can do any kind of evolution, and advance.

The Myth of Science

Any valid science is focused upon dealing with the dynamic patterns of living, and not with what these patterns have brought about. From this holistic perspective, our modern Western science is neither a science nor is it modern. By contrast, the I Ching that is five thousand years old, is both a true science and modern. It is a true scientific tool because it deals with the dynamic patterns of life and helps to identify them. And it is modern because its wisdom is without age, and still today fresh and new.

With modern science it's as with modern medicine. It deals with the symptoms and not the cause of a disease. It looks at secondary effects instead of regarding the primary causes. Instead of understanding that life is coded in dynamic patterns, it has assumed that life was founded upon certain principles. What is the difference between patterns and principles?

PROCESSED REALITY

A pattern is a set of things, a certain systemic arrangement I can make out in the complex scheme of reality. It is something I can observe. A pattern can be fix or it can be changeable. It can be static or dynamic. By contrast, a principle typically is the beginning of a down–hierarchy, a top–something in a kind of up–to–down order. It is not something I can directly observe. It's but the outcome of a conclusion I draw intellectually after observing nature. A principle thus contains my observer point or my judgment about reality.

Death science looks at life through the glasses of principles it has set before it was going to observe. It is essentially blind, and it proceeds by imposing characteristics upon nature. Western science is death science. Traditionally, it has gained its first conclusions about life by vivisecting cadavers, not by observing the moving changes of living. It is, and remains, a cadaver science that is far removed from the changing patterns of reality.

Life science looks at life without any prefixed principles or assumptions and observes the dynamic patterns or changes in the texture of life. It is a science that since its start, around five thousand years ago, was interested in life, and draw conclusions from life, and not from death. Traditional Eastern science is life science, one branch of this very large body of science and philosophy being Feng Shui. The I Ching is based upon life science, and is perhaps the highest condensation of it. Needless to add that, as such, it is non–judgmental and thus bears no moralistic judgments about human behavior. It looks at human behavior in exactly the same way it looks at all life patterns, and sees the changing nature of it before all.

Western scientific thought and traditional philosophy, ignorant about the fundamentals of the truth of the bioenergy as the primary creator force in the universe, was of course taking as real what it saw. As it saw matter only, it deducted that matter was the ultimate creator force, hence it assumed and formulated a basically materialistic scientific paradigm. However, this statement is valid only for mainstream science. As I have shown in various publications, even in times of the most fundamental repression of holistic pro–life wisdom in Europe, the original holistic life–science was taught and practiced in the underground by many alchemists and natural healers such as Paracelsus, to name only the most famous among them.

Today, mainstream science is like a lazy schoolboy, timidly learning lessons in dim afternoon classes it should have learnt, long ago, in the bright morning hours.

Chapter Four

Creating Reality

'Move on and leave your past behind you!,' said the wizard, and you answer, 'Yes, this is all right and good, and I can see it with my rational mind, but the rest of myself does not seem to follow that insight. So I remain caught in the net woven by my past and my deeply ingrained habitual thought and emotional patterns.'

I have the habit to make myself down, says one. I have the habit to make myself up, says the other. They talk to each other and conclude they were opposite characters. In reality, they are very similar. Both make themselves down. The second one however has a narcissistic pattern in addition to his guilt–and–shame pattern which means that he covers his wound by counterfeiting his own knowledge, saying I'm Peter Pan, and as such far removed from your petty world. I fly in the airs, catch me if you can. The sane mind does not make itself down, nor up. It accepts itself and all–that–is. His reality is either perceived directly and without distortion, which is possible and to be found with spiritual coaches and shamans, or processed to a minimum extent because of a highly developed consciousness surface, which is a state of spiritual evolution each of us can attain.

The present second part of this essay is a short guide to assist you in this very important quest for self–development. We have seen in the first part of this production that most people are on the passive side of life, so to say, perceiving reality more or less unconsciously and processing that information as good as they can. Perhaps it is true what David Mahoney who was named in Fortune Magazine one of the ten toughest bosses in America, has to say about this subject:

> I just keep moving every day as hard and fast as I can. High–intensity and high–voltage. Light comes from that, not from passivity. I insist we all do our

> best every day. I'm intense in everything I do and I expect others will be, too. There may be timing factors in it, good luck and fortune factors, but the question is, do you utilize it? Some of it you can't control—some of it goes against you—it works both ways. You run to daylight—where you see the break you go. Most people aren't even aware of what's happening around them. Two-thirds of the people don't know what's going on to them, personally.

I find it always amazing to see in which precise ways the insights of spiritual teachers, successful psychiatrists, famous artists and outstanding entrepreneurs coincide when it goes to explain the why and how of success. This tells me that this information and insight is available to all of us, and not only to some chosen elite. This insight is intuition. What most mediocre people do is to foreclose, in one or the other way, this natural knowledge about high achievement in order to justify their limitative worldview and to have a reason for engaging in self-pity and endless procrastination.

In all great success there is an element of novelty, something that was hardly predictable before the person succeeded on their particular path. This element of novelty is what makes the essential part of success in that it is part of a new reality that has been created, consciously or implicitly. As Edward de Bono states it:

—Once a new idea springs into existence it cannot be unthought.

No, I'm not talking about science fiction here. It's true that science fiction authors have been particularly imaginative for envisioning a new global reality, a new reality for the whole of humanity, and this particularly on a technological level. Here I'm talking about personal reality, not about the reality of a future humanity. I am not a science fiction author, nor a social utopist.

What I show you is not something related to myself, but something that is within your own personal potential. I show you an ability that you already possess, alongside your other skills and capacities. However, most people ignore that human imagination could have such a strong impact upon reality, and that it's actually a creator force, the creator force in the universe. Yet this tremendous energy has to be properly channeled. Your best imagination is not of much use when your general thought patterns are overwhelmingly negative.

Look at the life story of the great French novelist Honoré de Balzac, who was one of the most imaginative authors in the literary history of humanity. And yet his personal life was a series of tragedies, failures, disasters, scandals, open or hidden fights with others, animosities of the worst sort, and on the other hand unbridled debauchery, self indulgence and a lifestyle in which he exhibited very little self-discipline.

Suffices to read one page of this literary genius, the description of a person, the way the hero or heroine is clothed, walks, talks, thinks and we are put directly on-stage, facing that person in real life, so vivid are Balzac's descriptions, so brilliant and sharp was his imagination. But to what purpose was it used? It was certainly used to create great literature and art. It was hardly ever, or not at all used to create a new and different personal reality for the author himself.

This is an example for the fact that imagination alone does not bring the result, but that all depends on how imagination is channeled. How do you use your imagination? And with which purpose do you use it, when you use it?

When your memory surface is not clear, what happens when you use imagination for achieving your goals? I cannot tell you what happens, but I can tell you that chances are low for what you wish to happen really to come about. Why? Because your memory surface intermittently infiltrates information in your imaginative content that you absolutely do not wish to have put. Is there any willful control over this process? No. There is only one way: clearing the memory surface. When your glasses are dirty, and you see a foggy world, your willpower alone will not clean them. You have to take a piece of cloth and wipe them clean. It's the same with memory. It can be wiped clean.

I have done that at several instances in my life and thus I know that it works. You may have read in other books that it does not work, or that it works only for very exceptional people such as yogis, gurus, spiritual teachers, and the like.

No, we are talking here about something ordinary, not about a mysterious spiritual matter. We are talking about something rather mechanical. The memory surface pretty much works like a magnetic tape. It can store information. It can add–on information. It can erase information. Only one thing it cannot do. There is no function that stops the brain from recording.

This means that even though you may already program your reality according to your innermost wishes, reality always brings novelty of its own, because it's not, and cannot be, dependent upon your creative mind. That's a truth that our great poets express beautifully and that can be put very simply in the formula: reality always surpasses the individual mind. But that's not something to deplore. It only shows that our individual mind and soul are imbedded in a greater soul reality that kind of connects all minds within a cosmic meta–reality that is beyond the control of our individual mind. And yet, every impact of our individual mind upon this cosmic reality surface is noticed and can be retraced.

I have explained this in order to prevent you from getting depressed by just another pitfall of perception, this time of self–perception: the pitfall namely to believe we were insignificant as individual human beings on that cosmic, universal plane of consciousness. If this was so, we could not be co-creators, and we could not create our own reality. And in that case, I would not have taken the time and done the effort to tell you all of this.

A philosopher once compared humans with the billions of grains of sand on a beach, and this image has been interpreted as a metaphor for the insignificance of human beings in the cosmos. This is, in my view, a fundamental error.

Who, tell me, knows about the importance of an individual grain of sand in the whole of the cosmos, or even the whole of creation? To arrogate yourself to state that a grain of sand is insignificant means the same as saying that the whole of creation, and implicitly also that the creator force itself is insignificant. Today we know through quantum physics that every single electron, every single particle, that is only a tiny, very tiny fraction of a grain of sand, is conscious, and maintains relationships, chooses partners and friends, and locations, or remains undecided and at many locations at the same time.

Particles are conscious. And if that is true, by implication, grains of sand are conscious. And thus they are alive!

There are various methods to clear the memory surface. In order to make a good choice, you need to know more details about what memory actually is. Forget what you heard in school about it as it's most probably wrong. Memory is not in the brain, but in the luminous body or aura. It's coded in energy patterns and these patterns are virtually flowing around you, they are in movement, not static. The brain acts as interface to the memory surface; it does not store information. The old scientific view of the brain as a storage house is long superseded by newest research that, eventually, has included the insights we gain from parapsychology, clairvoyant research, Chinese medicine and acupuncture, and meditation, as well as quantum physics.

To conclude from this research, we can say that memory is volatile; this has, by the way, a big advantage, namely, that memory is not forever engraved anywhere in our gray matter, as it was believed by a mechanistic neurology of the 1960s and

70s. It also means that memory can be triggered to release information by touching parts of the body, by doing certain movements, by doing body work such as Rolfing or Alexander Technique. Reichian massage has proven to be especially conducive to releasing old memory patterns from the orgone shell or aura that permeates our organism, both inside the cell and around our physical body. I have also analyzed more recent techniques like Dr. Villoldo's soul retrieval and reviewed some of his books.

The second important point to know about memory is that it's not memory itself that creates hangups, addictions, habits or obsessions we may suffer from, but the emotional entanglement with past events and hurts that is a typical side–effect of trauma and abuse. It is entanglement that makes us repeat again and again the same scenarios in life, as our inner intelligence puts them on stage for us to get out of the strings, and heal our past.

The vicious circle in this is that when people are unconscious and blame life, god or others for their misfortunes, they are blocking the potential healing of their scars. Then they remain entangled, and perhaps so for their whole lifetime. That is why emonic consciousness is so important; it is energy consciousness, an awareness of the flow of energy in our organism, which includes awareness of where and how our energy flow is blocked or obstructed in certain parts of the body. Typically, it's the parts of the body that have been concerned when the abuse or traumatic event happened.

Now, how to build emonic or emotional awareness? The paradox that I found is that there is no technique to bring that awareness about when using our rational mind; it has to be built unconsciously, by sharpening our intuition.

How, then? The old Chinese saying Nonaction is action says it all in a way, when it's understood what this saying means. My experience with healing has taught me that it means to not directly interfere in the process, as this may strengthen the evil, so to speak. Let me give you one example for this from the book Getting Well Again (1978/1992) by the physician Dr. Carl Simonton. Dr. Simonton, who runs one of the most successful alternative therapies for cancer in California, reports in his book that many cancer patients who go to energy healers or laying–on of hands practitioners experience their cancer to grow, and not to shrink, after the treatment.

Why? Dr. Simonton says that cancer cells are very eager to receive energy, which lets them grow even more. This is an example that shows that a direct interference in the disease pattern does often not bring relief. And by the way, an operation, for example the removal of a cancerous tumor, is just another of these direct interventions; and it has been reported, by Simonton, and others, that removing a cancerous tumor does not per se remove the cancer, as the cancer is not in the tumor.

The tumor is only a secondary effect, one of many, of the cancer. This is why many cancer patients have made the sad experience that after having suffered a severe removal or amputation of an organ or limb, the cancer was beginning to spread elsewhere in the body.

So let me take this as a metaphor for introducing the simple yet effective technique I came to use for coping with hurtful memories; and let me add also that the expression erasing the memory surface is of course a metaphor as well. The process is much more complex in reality.

The technique I found helpful and effective for healing early trauma is creative writing. I came to realize it during a hypnotherapy twenty years ago, when my psychiatrist gave me certain themes to write about, asking me certain precise questions about my parents. I carried out these assignments very seriously and meticulously, and made the amazing discovery, that later was confirmed by my psychiatrist, that the actual healing took place every time before I had the next session with my psychiatrist, and thus actually before I presented those memoirs to him. He would utter something like we would not need to do any work, and can just 'chat a little today', as the big change was obvious and could even be seen in my face.

This dumbfounded me at first, but I had to report that indeed every time I wrote one of those little stories, a great calm came over me, an inner peace I had not known before, and I felt very clearly the stream of hot vital energy flowing through my whole organism, while before I felt the energy was stuck in my lower legs and my pelvis region, which is why I had icy feet most of the time. That problem with icy feet that I had been suffering from since my late adolescence was completely solved after writing the stories, and did no more recur later on in life.

On the other hand I have to say that honestly the writing itself was most of the time not a very agreeable experience.

The jotting down of those hurtful events, or in case the memory was only scarcely intact, the whole scenery or taste of a certain period of my life, triggered rather unwelcome body reactions, like outbursts of heat, hot rage, strong sweating, or sexual arousal, or all of this at once in a frenzy Draculian bath of violence that I can only compare with the eruption of a volcano. At other times the body seemed to shrink and mourn, and I felt like a small fly in a universe of ice, where there are endless pathways in the dark, and icy chambers with rotten souls everywhere. Then I would fall in a deep depression and had suicidal ideas.

Both the violent reactions and the suicidal ones were even stronger when I did not only the writing, but also used spontaneous art for triggering the inner healing. That is why I suggest to beginners to not do both at the same time, at least not when you are alone and have no psychiatric support at your side.

If you are serious about creating your own reality, instead of consuming the infected reality of the bulk of unconscious road-runners that populate this globe for too long, you also have to start with Life Authoring. You may also want to begin practicing a body consciousness technique such as *Tai Chi Chuan*. Last not least the excellent movie 'What the Bleep Do We Know?!,0 Quantum Edition offers many viable suggestions and scientific corroboration of the possibility to create our own personal reality—for good! And when you look over the fence, and in the art world, you may realize that some artists have done extremely well in creating new art reality. Let me mention only Pablo Picasso and Svjatoslav Richter here as examples while there are of course many more, but I know these particularly well.

These great artists provide excellent examples for reality creation; they have not only revolutionized their specific branch of artistry, painting, and musical performance, respectively, but with their strong personalities they have coined, each, a grandiose universe.

Let me close this chapter with two quotes from the book *The Power of Your Subconscious Mind* by Dr. Joseph Murphy:

> Look around you. Wherever you live, whatever circle of society you are part of, you will notice that the vast majority of people lives in the world without. Those who are more enlightened, however, are intensely involved with the world within. They realize – as you will, too – that the world within creates the world without. Your thoughts, feelings, and visualized imagery are the organizing principles of your experience. The world within is the only creative power. Everything you find in your world of expression has been created by you in the inner world of your mind, whether consciously or unconsciously.

You must ask believing, if you are to receive. Your mind moves from the thought to the thing. Unless there is first an image in the mind, it cannot move, for there would be nothing for it to move toward. Your prayer, which is your mental act, must be accepted as an image in your mind before the power from your subconscious will play upon it and make it productive. You must reach a point of acceptance in your mind, an unqualified and undisputed state of agreement.

BIBLIOGRAPHY

Contextual Bibliography

Ariès, Philippe

Centuries of Childhood
New York: Vintage Books, 1962

Arntz, William & Chasse, Betsy

What the Bleep Do We Know
20th Century Fox, 2005 (DVD)

Down The Rabbit Hole Quantum Edition
20th Century Fox, 2006 (3 DVD Set)

Covitz, Joel

Emotional Child Abuse
The Family Curse
Boston: Sigo Press, 1986

DeMause, Lloyd

The History of Childhood
New York, 1974

Foundations of Psychohistory
New York: Creative Roots, 1982

Diamond, Stephen A., May, Rollo

Anger, Madness, and the Daimonic
The Psychological Genesis of Violence, Evil and Creativity
New York: State University of New York Press, 1999

DiCarlo, Russell E. (Ed.)

Towards A New World View
Conversations at the Leading Edge
Erie, PA: Epic Publishing, 1996

Eisler, Riane

The Chalice and the Blade
Our history, Our future
San Francisco: Harper & Row, 1995

Sacred Pleasure: Sex, Myth and the Politics of the Body
New Paths to Power and Love
San Francisco: Harper & Row, 1996

BIBLIOGRAPHY

The Partnership Way
NEW TOOLS FOR LIVING AND LEARNING
WITH DAVID LOYE
BRANDON, VT: HOLISTIC EDUCATION PRESS, 1998

The Real Wealth of Nations
CREATING A CARING ECONOMICS
SAN FRANCISCO: BERRETT-KOEHLER PUBLISHERS, 2008

ELLIS, HAVELOCK

Sexual Inversion
REPUBLISHED
NEW YORK: UNIVERSITY PRESS OF THE PACIFIC, 2001
ORIGINALLY PUBLISHED IN 1897

The Sexual Impulse in Women
REPUBLISHED
NEW YORK: UNIVERSITY PRESS OF THE PACIFIC, 2001
ORIGINALLY PUBLISHED IN 1903

The Dance of Life
NEW YORK: GREENWOOD PRESS REPRINT EDITION, 1973
ORIGINALLY PUBLISHED IN 1923

ERICKSON, MILTON H.

My Voice Will Go With You
THE TEACHING TALES OF MILTON H. ERICKSON
BY SIDNEY ROSEN (ED.)
NEW YORK: NORTON & CO., 1991

Complete Works 1.0, CD-ROM
NEW YORK: MILTON H. ERICKSON FOUNDATION, 2001

Freud, Sigmund

The Interpretation of Dreams
New York: Avon, Reissue Edition, 1980
And in: The Standard Edition of the Complete Psychological Works of Sigmund Freud, (24 Volumes) ed. by James Strachey
New York: W. W. Norton & Company, 1976

Totem and Taboo
New York: Routledge, 1999
Originally published in 1913

Fromm, Erich

The Anatomy of Human Destructiveness
New York: Owl Book, 1992
Originally published in 1973

Escape from Freedom
New York: Owl Books, 1994
Originally published in 1941

To Have or To Be
New York: Continuum International Publishing, 1996
Originally published in 1976

The Art of Loving
New York: HarperPerennial, 2000
Originally published in 1956

BIBLIOGRAPHY

GOLEMAN, DANIEL

Emotional Intelligence
NEW YORK, BANTAM BOOKS, 1995

HAMEROFF, NEWBERG, WOOLF, BIERMAN

Consciousness
20 SCIENTISTS INTERVIEWED
DIRECTOR: GREGORY ALSBURY
5 DVD BOX SET, 540 MIN.
NEW YORK: ALSBURY FILMS, 2003

JAMES, WILLIAM

Writings 1902-1910
THE VARIETIES OF RELIGIOUS EXPERIENCE / PRAGMATISM / A PLURALISTIC UNIVERSE / THE MEANING OF TRUTH / SOME PROBLEMS OF PHILOSOPHY / ESSAYS
NEW YORK: LIBRARY OF AMERICA, 1988

JUNG, CARL GUSTAV

Archetypes of the Collective Unconscious
IN: THE BASIC WRITINGS OF C.G. JUNG
NEW YORK: THE MODERN LIBRARY, 1959, 358-407

Collected Works
NEW YORK, 1959

On the Nature of the Psyche
IN: THE BASIC WRITINGS OF C.G. JUNG
NEW YORK: THE MODERN LIBRARY, 1959, 47-133

Psychological Types
COLLECTED WRITINGS, VOL. 6
PRINCETON: PRINCETON UNIVERSITY PRESS, 1971

Psychology and Religion
IN: THE BASIC WRITINGS OF C.G. JUNG
NEW YORK: THE MODERN LIBRARY, 1959, 582-655

Religious and Psychological Problems of Alchemy
IN: THE BASIC WRITINGS OF C.G. JUNG
NEW YORK: THE MODERN LIBRARY, 1959, 537-581

The Basic Writings of C.G. Jung
NEW YORK: THE MODERN LIBRARY, 1959

The Development of Personality
COLLECTED WRITINGS, VOL. 17
PRINCETON: PRINCETON UNIVERSITY PRESS, 1954

The Meaning and Significance of Dreams
BOSTON: SIGO PRESS, 1991

The Myth of the Divine Child
IN: ESSAYS ON A SCIENCE OF MYTHOLOGY
PRINCETON, N.J.: PRINCETON UNIVERSITY PRESS BOLLINGEN SERIES XXII, 1969. (WITH KARL KERENYI)

Two Essays on Analytical Psychology
COLLECTED WRITINGS, VOL. 7
PRINCETON: PRINCETON UNIVERSITY PRESS, 1972
FIRST PUBLISHED BY ROUTLEDGE & KEGAN PAUL, LTD., 1953

BIBLIOGRAPHY

KOESTLER, ARTHUR

The Act of Creation
NEW YORK: PENGUIN ARKANA, 1989.
ORIGINALLY PUBLISHED IN 1964

KRISHNAMURTI, J.

Freedom From The Known
SAN FRANCISCO: HARPER & ROW, 1969

The First and Last Freedom
SAN FRANCISCO: HARPER & ROW, 1975

Education and the Significance of Life
LONDON: VICTOR GOLLANCZ, 1978

Commentaries on Living
FIRST SERIES
LONDON: VICTOR GOLLANCZ, 1985

Commentaries on Living
SECOND SERIES
LONDON: VICTOR GOLLANCZ, 1986

Krishnamurti's Journal
LONDON: VICTOR GOLLANCZ, 1987

Krishnamurti's Notebook
LONDON: VICTOR GOLLANCZ, 1986

Beyond Violence
LONDON: VICTOR GOLLANCZ, 1985

Beginnings of Learning
NEW YORK: PENGUIN, 1986

The Penguin Krishnamurti Reader
NEW YORK: PENGUIN, 1987

On God
SAN FRANCISCO: HARPER & ROW, 1992

On Fear
SAN FRANCISCO: HARPER & ROW, 1995

The Essential Krishnamurti
SAN FRANCISCO: HARPER & ROW, 1996

The Ending of Time
WITH DR. DAVID BOHM
SAN FRANCISCO: HARPER & ROW, 1985

LIEDLOFF, JEAN

Continuum Concept
IN SEARCH OF HAPPINESS LOST
NEW YORK: PERSEUS BOOKS, 1986
FIRST PUBLISHED IN 1977

MOORE, THOMAS

Care of the Soul
A GUIDE FOR CULTIVATING DEPTH AND SACREDNESS IN EVERYDAY LIFE
NEW YORK: HARPER & COLLINS, 1994

BIBLIOGRAPHY

Rosen, Sydney (Ed.)

My Voice Will Go With You
The Teaching Tales of Milton H. Erickson
New York: Norton & Co., 1991 Stein, Robert M.

Redeeming the Inner Child in Marriage and Therapy
In: Reclaiming the Inner Child
Ed. by Jeremiah Abrams
New York: Tarcher/Putnam, 1990, 261 ff.

Steiner, Rudolf

Theosophy
An Introduction to the Spiritual Processes in Human Life and in the Cosmos
New York: Anthroposophic Press, 1994

Stone, Hal & Stone, Sidra

Embracing Our Selves
The Voice Dialogue Manual
San Rafael, CA: New World Library, 1989

Szasz, Thomas

The Myth of Mental Illness
New York: Harper & Row, 1984

Tart, Charles T.

Altered States of Consciousness
A BOOK OF READINGS
HOBOKEN, N.J.: WILEY & SONS, 1969

What the Bleep Do We Know!?

See Arntz, William

Whitfield, Charles L.

Healing the Child Within
DEERFIELD BEACH, FL: HEALTH COMMUNICATIONS, 1987

Personal Notes

www.ingramcontent.com/pod-product-compliance
Lightning Source LLC
Chambersburg PA
CBHW020431220526
45464CB00002B/653